DATE DUE

by Victor Gentle and Janet Perry

Gareth Stevens Publishing
A WORLD ALMANAC EDUCATION GROUP COMPANY

Please visit our web site at: www.garethstevens.com
For a free color catalog describing Gareth Stevens' list of high-quality books and multimedia programs, call 1-800-542-2595 (USA) or 1-800-461-9120 (Canada). Gareth Stevens Publishing's Fax: (414) 332-3567.

Library of Congress Cataloging-in-Publication Data

Gentle, Victor.
Fires / by Victor Gentle and Janet Perry.
p. cm. — (Natural disasters: an imagination library series)
Includes bibliographical references and index.
ISBN 0-8368-2833-X (lib. bdg.)
1. Wildfires—Juvenile literature. [1. Wildfires. 2. Forest fires.
3. Fires.] I. Perry, Janet, 1960- II. Title. III. Series.
SD421.23.G55 2001
363.37'9—dc21 00-051620

First published in 2001 by
Gareth Stevens Publishing
A World Almanac Education Group Company
330 West Olive Street, Suite 100
Milwaukee, WI 53212 USA

Text: Victor Gentle and Janet Perry
Page layout: Victor Gentle, Janet Perry, and Joel Bucaro
Cover design: Joel Bucaro
Series editors: Mary Dykstra, Katherine Meitner
Picture researcher: Diane Laska-Swanke

Photo credits: Cover © G. Perkins/Visuals Unlimited; p. 5 © Peter K. Ziminski/Visuals Unlimited; p. 7 © Joe McDonald/Visuals Unlimited; p. 9 © Mark Downey/Lucid Images; pp. 11, 13 © Raymond Gehman/CORBIS; p. 15 © Richard Olsenius/NGS Image Collection; pp. 17, 21 © AP/Wide World Photos; p. 19 © C. Andrew Henley/Larus

Printed in the United States of America

1 2 3 4 5 6 7 8 9 05 04 03 02 01

Front cover: *Very few forest animals die from wildfire. This doe and fawn can easily escape danger.*

TABLE OF CONTENTS

Words that appear in the glossary are printed in **boldface** type the first time they occur in the text.

Don't Play with Fire!

When you were little, grown-ups told you not to play with fire. *Someone* could get hurt.

Fire is dangerous, that's true. It destroys places and kills living things. People need to be careful about how they use it.

People also should be careful not to get in nature's way when fire is involved in "housecleaning." Fires clean up prairies, forests, and swamps by clearing out dead plants. Then new plants can grow in their place. Some animals are driven away by fire to find new homes. For other animals, a **fire season** is a normal part of the weather. These animals depend on nature to clear up their homes this way.

Elk and pines return after the 1988 Yellowstone National Park wildfire. Experts say that more kinds of plants grow after wildfires, improving the ***ecosystem****.*

Fire Weather

Wildfires are uncontrolled fires that burn in the wilderness. They happen each year during the fire season in places with hot, dry **climates**. These places always have wildfires during their hot season.

In parts of the western United States and the Sierras of Mexico, a fire season is part of the normal climate. Most of Italy, the southern coasts of Spain and France, and Australia's bush areas also have fire seasons.

Certain weather conditions, such as long periods of time without rain, make it easy for wildfires to start — even in areas without a fire season. Wildfires also burn in wet areas, such as Florida's swamps — or in cold places, such as Russia's forests, too.

Wildfire in the swampy Everglades burns cypress trees. Surface fire sweeps the grasses and travels up tree trunks to treetops, where it leaps from tree to tree as a crown fire.

LIVE FIRE

Like you, fire needs air, fuel, and heat to live. If any of these three things is missing, there is no fire.

Air, fuel, and heat form what is called the **fire triangle**.

You breathe air. Your body gets energy from the food you eat. If you don't stay warm enough, your body will stop working.

A fire also "breathes" the air around it and continues to burn by "eating" **combustible** materials, such as dry grasses, wood, leaves, gasoline, and trash. If the fire is cooled by water or air, cannot get air, or runs out of fuel, it dies.

A ***pyrocumulous*** *cloud is formed by the steam and heat of this fire in Oakland, California. The storm's lightning strikes may have started more fires.*

WILDFIRE, TO THE RESCUE!

Swamps, grasslands, and forests are all ecosystems that need fire to stay healthy. An ecosystem is a place with a particular climate where a specific group of **organisms** live together.

The organisms in an ecosystem are not very careful about "cleaning house." Animals and plants litter the wilderness with their waste and dead bodies. The living plants are choked by the mess, and new plants have no place to grow. Soon, there is little for the animals to eat.

That's when weather can come to the rescue. A lightning strike can cause a wildfire, which burns up all the trash and makes room for new plants.

This photo shows how a ***prescribed burn*** *improved the health of a forest in Idaho. The green patches were treated with fire, the brown patches were not.*

Who Can Stand the Heat?

Sometimes organisms invade an ecosystem and challenge the plants and animals that already live there. But some of these invaders cannot adapt to the wildfires that naturally occur in these ecosystems. Fire either destroys the newcomers or drives them out.

Some organisms that are native to ecosystems with a fire season can't live without wildfire. Lodgepole pinecones won't release their seeds until they are heated to 113° F (45° C). Grasses in swamps and prairies can regrow from their **corms**, which are safe underground. Animals like prairie dogs and elk can eat the plants and insects that survive a fire.

Bison nibble new plants that sprouted out of the freshly burned surface of the forest. The burn clears the forest floor, allowing more plants to grow than before.

Fire Works!

Many peoples have used fire to improve the land where they live. Natives to grasslands all over the world noticed that the animals they hunted grazed in large numbers on the new grasses that sprouted after a prairie fire. Ranchers today still use fire to improve the grasslands for their herds of cattle. Rangers in wilderness areas also start fires to take care of forests, grasslands, and marshes.

These are called prescribed burns. Besides protecting ecosystems in wilderness areas, prescribed burns protect the homes of people who live nearby. By clearing out dead plants, prescribed burns keep combustible materials available to wildfires at a safe level. Otherwise, when wildfires occur, they may rage out of control.

At the Z-Bar Ranch in Strong City, Kansas, burning the grasslands means new grass for the cattle to eat.

FIGHT FIRE WITH FIRE

Do people plan every fire they start? No! Most fires are accidents that burn out of control. Maybe someone forgot to put out a campfire, left an iron on, or played with matches.

Fires get worse if they are blown around by the wind, if they are fed by chemicals (such as gasoline), or if there is fuel (like wooden houses, dry meadows, or forests) in the fire's path.

Even fires that start out as prescribed burns can become uncontrolled blazes that destroy hundreds of thousands of acres (hectares) of plants, animals, people, and property.

By carefully burning a section of forest ahead of a blaze, firefighters made a ***firebreak****. The blaze stops here, because there is nothing left to feed the fire.*

Watch Out for That Tree!

History shows us that people must be careful how they plant, build, and burn. The Great London Fire in 1666 turned the city into a raging blaze in just eight hours. Many of the buildings in London were made of wood, with roofs made of thick grasses called reeds. The fire jumped from roof to roof, burning buildings to the ground. Most of London was ashes in three days. Although 100,000 people lost their homes, only six died.

The 1991 Oakland Hills fire in California was even worse than the London Fire. Extremely fast winds, dry weather, a lack of water, and "fuel" from the eucalyptus trees made a hot, fast blaze. The fire burned 3,469 homes, killed twenty-five people, and caused $1.5 million in damage.

This eucalyptus tree is burning at its roots in a ground fire. These trees have oils in them that also burn quickly to the treetops, throwing sparks to other trees — a crown fire.

Fire Safe

What can we do to prevent wildfires from causing disasters?

We can build houses that are not easily burned, with designs that prevent fire from jumping from roof to roof. We can plant fire-resistant trees and bushes.

By clearing away trees and brush around our homes, we can build firebreaks. We can also make sure that fires that we start are completely out before we leave them.

Playing with fire is a very bad idea. A very good idea is to understand how fire is made, how it works, and how to use it safely as a tool.

Malibu, California, laws require fire prevention. The hill on which this house sits has been cleared of dry brush that could possibly fuel a wildfire.

More to Read and View

Books (Nonfiction)

Atlas of the Earth. Alexa Stace (Gareth Stevens)
Disaster! (Catastrophes That Shook the World). Richard Bonson and Richard Platt (Dorling Kindersley)
Endangered Environments! Bob Burton (Gareth Stevens)
Fire and Flood. Nicola Barber (Barron's)
Fire in Oakland, California: Billion Dollar Blaze. Carmen Bredeson (Enslow)
Lightning. Stephen Kramer (Carolrhoda)
The Science of Air. Sarah Dunn (Gareth Stevens)
True Disaster Stories. Karen Connor O'Sweeney (Franklin Watts)
Whatever the Weather. Karen Wallace (Dorling Kindersley)
Wild Weather (Associated Press Library of Disasters). Matt Levine, Editor (Grolier)
Wildfire. Patrick Cone (Carolrhoda)

Books (Activity)

Trees (series). Peter Mellett (Gareth Stevens)
Weather (series). Robin Kerrod (Gareth Stevens)

Videos (Nonfiction)

Chasing El Niño. (WGBH Boston Video)
Lightning! (WGBH Boston Video)
Weather Fundamentals: Climate and Seasons. (Schlessinger Media)
Weather Fundamentals: Wind. (Schlessinger Media)

Web Sites

If you have your own computer and Internet access, great! If not, most libraries have Internet access. The Internet changes every day, and web sites come and go. We believe the sites we recommend here are likely to last and give the best and most appropriate links for our readers to pursue their interest in wildfires and the places they happen.

www.ajkids.com

Ask Jeeves Kids. This is a great research tool.
Some questions to try out in Ask Jeeves Kids:
How are most forest fires started?
What is a firestorm?

You can also just type in words and phrases with "?" at the end, for example:
Fireproof house?
Lodgepole pine?
Prescribed burn?

www.smokeybear.com

Visit Smokey Bear! Play some games, find pictures to color, and see how you can help Smokey prevent forest fires.

www.usfa.fema.gov/kids/

This is the United States Forest Service's kids' page, where you can become a junior fire marshal, discover how to plan an escape from a fire, visit Hydro's Hazard House, and play more games that make fire safety fun!

www.conwayfiremuseum.org

At the Conway Fire Museum, you can click on the Exhibits button to see antique fire engines and read about how they operated.

www.wildfiremagazine.com/kids.shtml

At Wildfire Magazine, you can go to the Kids page and click on many different buttons to go right inside the world of firefighting and fire prevention!

Click on:

Fire Prevention: Meet "The Preventor" and read his story! Get coloring pages, crossword puzzles, and word searches.

Yellowstone Fire of 1988: Read the story of the entire fire at Yellowstone in a series of news articles.

How a Fire is Fought: Read about the stages of a fire and the ways that firefighters control a fire.

www.chicagohs.org/fire/

Go to the Great Chicago Fire to read how the Great Chicago Fire burned the city down. Click on the special media button to hear songs, see the fire burning, and get a panoramic image.

Note: For this to work, you need to have Quicktime and Shockwave software, plus 3-D glasses.

Glossary

You can find these words on the pages listed. Reading a word in a sentence helps you understand it even better.

climate (KLI-met) — a pattern of weather that is common to a particular place 6

combustible (kom-BUS-tah-bul) — able to burn 8

corms (kormz) — the fleshy, thick roots of plants like lilies, grasses and carrots 12

ecosystem (E-ko-sis-tum) — a place that has a certain weather pattern and has specific organisms that live together, adapted to the place, its climate, and each other 4, 10, 12, 14

firebreaks (FIRE-brakes) — sections of land that are cleared by fire, water, or sand so that the fuel for a fire is smothered, cooled, or destroyed 16, 20

fire season — a section of time during which a type of ecosystem will withstand burning 4

fire triangle (fire TRI-ane-gul) — three things a fire needs to live: air, fuel, and heat 8

organism (OR-gun-is-um)— living things, such as animals, plants, birds, bugs, germs, and fungus 10, 12

prescribed burn (pre-SKRIBE-d burn) — a fire set on purpose to clear land, so new things can grow and trash is destroyed 10, 14, 16

pyrocumulous (pi-ro-KYOOM-u-lus) — a thunderstorm cloud created by the steam and heat of a wildfire 8

Index